I0493526

CONTENTS

INTRODUCTION

Colours have always fascinated man and right from the pre historic times he has always tried to use colour in whichever way possible in his life. Initially they were extracted from nature (i.e., from flowers, fruits) and these colours were known as the natural colours. He had been using various means to decorate the clothes he wore and here the chief source of colouring matter, until about a century ago, was natural colours especially the ones obtained from nature. These natural colours were initially used to dye natural fibres like cotton, wool, linen, silk etc. With the advent of industrialization man started dyeing clothes on a commercial basis and then his main aim was to find a way to increase the production at the least expense. It was soon realized that the method of extraction of natural dyes and their application were lengthy and laborious. This gave the impetus to development of synthetic dyestuffs.

The introduction of synthetic colours was a big boon to the colourists and hence it was accepted widely and practiced extensively. Though these dyes greatly reduced the time and there by increased the production they, had a lot of disadvantages. Most of the synthetic dyes cause ecological concerns unlike the natural dyes, which were eco-friendly. This led to the development of innovative dyeing techniques in an attempt to make the dyeing procedure more eco-friendly and also to reduce the cost, time and energy involved in the dyeing of textile material using synthetic dyes.

For further developments in textile colouring it was important for man to have knowledge on compatibility of dyes. Knowledge in this field facilitated invention of millions of shades and there by increasing the spectrum of colours that could be used in dyeing of textile materials. But all this developments had one common goal and that was to produce coloured textile of desired shade, homogeneous depth and hue with due consideration to economy and ecology.

DEVELOPMENTS IN DYEING TECHNIQUES

1) REACTIVE DYES

1.1) General dyeing process of reactive dyes:

The exhaustion of reactive on to the fibre involves two stages[6].

This is primary exhaustion stage

The first being physical adsorption as a result of substantivity that occurs by addition of dye and inorganic salt.

This is secondary exhaustion stage

In the second stage dye absorption and fixation takes place as a result of addition of alkali to the bath.

1.2) Positives and negatives

These dyes offers several advantages such as low cost, ease of application, wide gamut of colours and good fastness to light, washing and rubbing. However, in case of reactive dyes, high concentrations of salt are required during dyeing for effective exhaustion of dye onto the cotton surface. Although the use of salt is economical, it increases the total dissolved solids in the effluent stream. Further high electorate concentrations of dye-bath discharges are undesirable as increased salinity in rivers upsets the delicate balance of flora and fauna[8].

There are many parameters that control dyeing of reactive dyes and one of the main factors that affect the substantivity of the dyes is the addition of auxiliaries and other Speciality chemicals which will help in the dye uptake. The developments in dyeing techniques of reactive dyes has mainly been to reduce the auxiliaries so as to reduce the cost as well as the to decrease the effluents.

The dye hydrolysis can be minimized using controlled colouration technique like low liquor ratio machines, different alkali and electrolyte systems and Speciality chemicals to increase dye uptake.

So there were many developments in dyeing procedures made to overcome the negatives. Few of the developments or modifications in the dyeing procedures are given below.

1.3) Modification in dyeing techniques of reactive dyes

- **Low salt dyeing**

Driving force behind this is the environmental concern. With increase in reactive dyes it has become necessary to counter the damage done by reactive dyeing. Reduced quantity of salt enables optimum method of addition of salt to be selected. In critical cases where liquid salt addition is necessary lower quantity can be made possible without altering bath volume greatly. This saves time and assists level dyeing. Easier dyeing also becomes easier to wash off in comparison to dyeings using conventional quantity of salt. Saving salt between 30-50% is possible[7].

- **Replacement of salt by speciality product**

There were many studies conducted to find the possibility for the replacement of salt by an eco friendly product. An eco salt was developed which could be used for the dyeing of cotton fabrics with hetero-bifunctional reactive colours.

Conventional dyeing method of reactive dye (i.e. addition of common salt for exhaustion) was compared with a dyeing method in which the common salt was replaced by the eco-salt. A dyeing procedure where both exhaustion and fixation are made to take place simultaneously was also carried out. There was also the all- in method where both dyes, speciality product and common salt were added together.

Now the results showed that, exhaustion of dyebath in the presence of salt or the speciality product are more or less the same when dyeing was carried out by conventional or all- in one method. Similarly fixation was also found to be same at both 2% and 4% depth of shade. It was found that the fixation of the dye increased when the concentration of the eco-salt was increased from 10 gpl to 20 gpl as compared with the common salt also the colour value was found to be higher in the all in one method when eco-salt was used.

Hence from the tests conducted the it can be concluded that it is possible to replace the conventionally used salt by eco-salt E during dyeing of cotton with heterobifunctional; reactive dyes without affecting the colour value as well as fastness characteristics of the dyes. Also the product is biodegradable and decreases TDS in the effluent stream.

- **Salts**

Fibre reactive dyes require higher quantity of salt than other cellulosic dyes. Some plants have increased the general recommendations from dye manufacture to as much as 20-30%. Some plants use up to 100-150gpl. But the environmental concerns have required plants to reduce salt in the effluents. Industry has moved from 15:1-20:1 to 8:1-5:1 ratios. This improves the substantivity of fibre reactive dyes thus reducing the quantity of salt required. Today there are products that require only 30-50gpl at 10:1 ratios for optimum yield. A computer program has been devised to enable the dyer to adjust dye to salt ratio to varying dyeing conditions and to calculate quantity of alkali with stored data computer provides formulation for salt and MLR.

- **Electrolytes**

NaCl is the most efficient salt for short dyeing time of 20 mins. While LiCl yields higher dye uptake during prolonged dyeing periods of 60mins. Effect of electrolyte on dye uptake is related to radius of solvated cation. Radius of $CsCl < KCl < Na_2SO_4 < NaCl < LiCl$.[4] Dye uptake increases with salt concentration except with LiCl, which shows maximum uptake of dye at concentration of 40gpl.

- **Novel methods of addition of alkali for better results**

Reactive dyes are fixed on cellulose by formation of covalent bonds between hydroxyl group of fibre substrate and functional group of dye. Fixation of these dyes is by sodium carbonate. To economise this process feasibility of using weak and strong alkali was attempted.

1. Conventional method: use 20 gpl sodium carbonate in two steps with interval of 15 mins. After 30 mins of exhaustion.

2. Alternative method: mixture of sodium carbonate and sodium hydroxide is used.

3. Modified method: sodium carbonate is added in steps while sodium hydroxide as after treatment.

Effect of concentration of sodium carbonate and sodium hydroxide was studied by varying sodium carbonate concentration (1-10 gpl) and sodium hydroxide (0.5 gpl) and also by varying sodium hydroxide concentration (0.1-2 gpl) while fixed sodium carbonate concentration (5 gpl).

It was observed that concentration of 5gpl sodium carbonate and 0.5 gpl sodium hydroxide gave the best results with respect to colour strength[9].

- **Use of sodium silicate**

Use of sodium silicate as fixation agent in place of sodium carbonate in reactive dyeing of cotton fabrics with Procion, Amaryl, Remazol dyes was found to improve dye fixation and economical. This was more beneficial for cold brand dyes rather than hot brand dyes.

- **Dyeing of silk in the presence of potassium periodate**

Silk fibre can be dyed with direct acid reactive and basic dyes. In most of the system, the dyeing takes place in the presence of an acid at boil and under this condition, it takes about an hour to get complete exhaustion of the dyebath. However, dyeing at boil damages the silk, deteriorating its qualities.

It became a necessity to find a technique of dyeing silk with low temperature so many experiments were carried out. Experiments were conducted in dyeing of silk with C.I.ReactiveYellow 14 and the effect of KIO4 on the above dyeing procedure was investigated.

Experiments were carried out with different concentrations of KIO4 and at different temperatures. By analyzing all the results it was concluded that silk fabrics could be dyed at low temperature with reactive dyes in the presence of KIO4. the optimum conditions for dyeing is, temperature of 50deg c, time 45 min and concentration of 0.015 mole/litre[8].

- **Wool dyeing with reactive dyes**

Reactive colours can be applied on cellulosics from an alkaline bath. Since protein fibres are sensitive to alkali they cannot be dyed under conditions employed for cellulosics requiring modified dyeing conditions. Wool is a protein fibre composed of amino acids containing basic amino and acidic carboxyl group which on condensation forms an amido group and finally polyamide macromolecule or polypeptide. Wool keratin contains variety of groups such as amino, hydroxyl and thiol that can combine with reactive dyes under acidic or neutral conditions with no danger of hydrolysis of reactive group. Due to strong covalent bond the dyes become integral part of wool keratin hence removal is difficult.

It was attempted to apply reactive colours with four classes of reactive systems on wool to increase dye uptake by partially reducing wool thereby increasing thiol groups which are the dyeing sites. Partial reduction was carried by using sodium sulphite, thioglycollic acid (TGA). The reduced wool was dyed with commercial dyes namely Procion Brilliant Red M8B, Procion Green HE4BD, Remazol Yellow GR and Drimalan Yellow F3GL with MLR 1:30 at ph 5 and 7 for 1 hour at boil. The concentration of stripped dye solution was determined spectrophotometrically.

Given below are the light fastness rating for wool and reduced wool when dyed with Procion Red M8B, Remazol Yellow GR, and Drimalan Yellow F3GL[7].

Reactive dyestuff	Light fastness ratings: Intact wool	Light fastness ratings: Reduced wool
Procion Red M8B	4	4-5
Remazol Yellow GR	4	4-5
Drimalan Yellow F3GL	4-5	4-5

The reduced wool was shown to give relatively good light fastness as shown in the tabular column.

- **Acid fixing reactive dyes**

Procion T dyes a range of liquid reactive dyes based on phosphonic acid reactive group are fixed under acidic conditions in presence of carboimide at around 200^0c. Large quantities of fixatives are needed. They do not hydrolyse at end of dyeing process and retain their capacity for fixation. High fixation values can be obtained using solvents and cynamide at 160^0c. Fixation efficiency can be improved by steaming after padding and drying.

- **Use of mixed urea-potassium thiocyanate plasticizer**

Use of mixed urea- potassium thiocyanate plasticizer system for thermo fixation of reactive dyes causes decrease in urea decomposition during heat treatment leading to 1.6 to2 times increase in reactive dye fixation and 15 to 20% increase in dyeability of cotton fabric. The mixed plasticizer creates more favourable conditions for interaction of dye with fibre due to plasticization of cellulose both structurally by urea and molecularly by potassium thiocyanate. When only urea is employed decrease in degree of dye fixation at temperature above 160^0c is observed due to lowering of ph of cured fabric by the thermal decomposition products of urea. This reduces the dye cotton covalent bonding. Addition of potassium thiocynate favours decrease in urea decomposition during curing hence lower decrease in pH.

2) ACID DYES

Acid dyes are anionic dyes characterized by substantivity for protein fibres like wool, silk and polyamide.

2.1) General dyeing process of acid dyes

The general dyeing procedure of dyeing of wool or other protein fibres with acid dyes involves the addition of acid to he dye bath along with dye and water. When wool is introduced to this dye bath the amino groups get protonated and there by causing an electrostatic attraction between the protonated group of the substrate and the –RSO3- of the dye and there are also chances of hydrogen bond formation between the hydroxy group of the dye molecule and the ketonic group of the substrate.

2.2) Positives and negatives

They are one of the best dyes to dye wool, other protein fibres and polyamide fibres and hence they are very largely used. They also produce wide range of brilliant shades.

The main disadvantage is the use of acid in the dyeing, hence the application of these dyes can degrade acid sensitive fibres and hence its use gets restricted and moreover the use of acid also causes effluent problems. It is also necessary to maintain the pH during the dyeing process inorder to get good dyeing results.

2.3) Modifications in the dyeing techniques of acid dyes

- **pH control for poly amide dye bath in a reused dye bath**

The main objective for controlling the pH through out the dyeing cycle acid dyeing is to permit the dyer to produce level, reproducible results but recycling of the dyebath can affect the dyebath.

The advantage of dyebath recycling is numerous and includes less water to heat, less effluent to treat and reduced consumption of energy, dyes and chemical auxiliaries. Because of the potential economic and environmental advantages involved in this process, clearly it deserves further investigation.

The use of buffer system requires a greater awareness not only of the factors that determine pH, but also that stabilize it against interference. A buffer system is usually a weak acid and its corresponding salt or a weak base and its salt.

A pH sliding system is particularly useful for non-migrating acid dyes on nylon. Using products that release acidic compounds as they undergo decomposition with increase in dyeing temperature can carry out the controlled lowering of the pH. Eg- ammonium salts (sulphate, acetate, tartrate) and the most widely used is ammonium sulphate.which decomposes giving ammonia (gas) and sulphuric acid. However in enclosed machines its not that efficient as ammonia is prevented from evaporating.

The use of hydrolysable esters dates back to 1953 when brotherton co. ltd introduced their estrocon process for which diethyl tartrate or ethyl lactate was used wool dyeing with acid milling dyes and chrome dyes by the single bath method.

In this study we report a number of pH control systems for dyeing polyamide fibres in reused dyebath. The dyeing properties and repeatability are compared with their ability to control the pH.

Experimentally acid dyeing of nylon was carried out with 100 ml dyebath, 3g sample and acid donors as shown in the table. The acid donors analysed are Sodium dihydrogen phosphate(SDP), Ammonium sulphate(AS), gamma-butryolactone(GBL), Ethyl lactate(EL), Diethyl tartrate(DET)[10].

PH control system	Acid donors	Concentration	Initial pH adjustment
Buffer system	Sodium dihydrogen phosphate(SDP)	3.5	6.5
pH sliding system	Ammonium sulphate(AS),	2.0	8.0
pH sliding system	gamma-butryolactone(GBL)	0.10	8.0
pH sliding system	Ethyl lactate(EL	0.10	8.0
pH sliding system	Diethyl tartrate(DET).	0.10	8.0

pH control systems used in the study

Dye bath reuse procedure-the dyebath is left to cool to RT then 20 ml of water with necessary amounts of dye, leveling agent and acid donors (to compesate for the losses by evaporation and fibre drag out. The initial pH of the dyebath containing SDP was not readjusted while others (hydrolysable esters) were readjusted to 8 by using sodium carbonate after each dyeing.

Dye characteristics and dyeing properties

Fastness

The fastness of the colour to washing, rubbing, perspiration was found for the dyed materials by ISO methods. Colourfastness figures were similar irrespective of the pH control system and the number of dye bath-reuse.

From the results it is found that no deterioration in colour fastness could be detected for the fabrics dyed during 10 cycles of the dye bath reuse.

pH variation: as the number of reuse cycle increases, the residual pH of the dye bath containing AS decreased slowly, but not below 8. Where as SDP and GBL the pH increased slightly from 6.57-6.93 and 5.51-5.85 respectively. In the case of EL and DET the pH was controlled in the range of 5.42-5.58 and 4.88-5.07 respectively.

% Exhaustion: the hydrolysable organic esters showed high and stable exhaustion, close to 100%, while AS showed relatively low and fluctuating values. SDP also showed a relatively good exhaustion values but decreases due to the limitations in the buffering capacity.

The CIE Lab a* and b* values of all the dyed samples were studied with number of dye bath reuse cycles and it was found that AS and SDP gave bluish colours than the hydrolysable organic esters.

Conductivity: conductivity is the extent of the salt present in the dye bath and higher conductivity owes to excess of salt aggregation in the bath and ultimately causing unevenness dyeing. AS gave higher conductive values as compared to the other hydrolysable esters. SDP also have high conductivity values but it is not dramatically higher than the hydrolysable esters.

Cost factors: the cost of GBL, EL, and DET relative to AS, which is commonly used in the dyeing of polyamide is 1.02, 1.13 and 12.0 respectively. Therefore, in the case of GBL and EL, the excellent pH control ability in the dye bath reuse system could offset the higher cost, especially when used in the closed system dyeing.

This study has demonstrated that the use of hydrolysable organic esters as acid donors for pH control offers distinct advantages in the dyeing of polyamide with a reused dye bath.

- **Low temperature dyeing of nylon by pH shift method**

The main objective for controlling the pH is to produce level and batch-to-batch reproducible results in the dyeing of polyamide fibres.

In this particular development we have considered dyeing of nylon 6,6 with the help of alkali as well as an acid. The main objective behind this development is to find if it is possible to reduce the time and temperature in the conventional dyeing method by introducing the pH controlled method[11].

Dyeing experiment is carried out both by the conventional method and by the new development.

Conventional method

Dyeing of nylon 6,6 is done from a dye bath containing the required amount of the dye wetting agent (0.4 gpl), non-ionic leveling agent (0.04 gpl) and acetic acid (pH 4.5) set at 60 deg c. MLR of 1:100 gradually raised to 95 deg c within 30 min and dyed at this temperature for 30 min.

pH controlled method

0.5 g of the sample is taken in a dye bath (1:100) with wetting agent (0.4 gpl) and the dye bath is preset with pH ranges from 11 to 8 (just to study the pH), this is done by adding alkali. After dyeing in the alkaline bath for 30 min, the pH of the dye bath was shifted to 4.5 with help of sulphuric acid (10 N). the dyeing was further carried on for 30 min in the constant pH 4.5.

Optical density of the dye bath is found at regular interval of 10 min during dyeing and the test samples are returned to the bath to maintain the liquor.

The amount of acid and alkali that is to be added is calculated by calibration in blank bath. Results given below

Amt of NaOH (2.5 N)	Desired pH	Amt of sulphuric acid (10 N)	Desired pH
0.3	11	Alkaline blank bath	11
0.2	10	0.02	10
0.16	9	0.03	9
0.12	8	0.04	8
Blank bath	7	0.05	7

There were three dyes that were considered for the experiment and they are Acid red 88, Acid red 14, Acid red 18. Acid Red 88 forms aggregates where as Acid Red 18 splits molecularly.Acid Red 88 is a mono sulphonic acid dye, Acid Red 18 is a tri sulphonic acid dye and Acid Red 14 is a di sulphonic acid dye.

Dye characteristics and dyeing properties

In the conventional method of dyeing there is steady reduction in the optical density of the dye bath for all the three dyes that were taken under consideration with increase in the dyeing time and reaches more or less constant value after almost 1 hr.

Whereas in the shift in pH method where dyeing is commenced at 60 deg c with various initial starting alkaline dye bath ranging from 8-11, the value of optical density increases gradually up to a period of 30 min dyeing time and the extent depends on the pH (i.e. the higher the alkalinity the higher the optical density)

This increase in the optical density is attributed to the change in the intensity of the dye due to change in the pH which causes a shift in adsorption to a region of a longer wavelength, the colour seen being 'deeper', a term denoting a change of the hue i.e. bathochromic effect.

The absorption was found to increase in the case of Acid Red 88, which contains one sulphonic acid group and has a tendency to get aggregated.

For acid red 18, which is molecularly dissolved because of the higher number of sulphonic acid groups shows decreased interaction.

Dyeability comparison

By studying the optical density data of the shift in pH method it was found that the nylon 6,6 cannot get dyed in the initial 30 minutes (alkaline dye bath) this could be attributed to the fact that nylon is not protonated in the alkaline medium and later when the pH was brought down to 4.5 by using sulphuric acid the exhaustion and dye uptake increased dramatically and the dyeing could be completed in 45 min instead of the usual 60 min.

The dyeability also depends on the number of sulphonoic acid groups in the dye molecule. That is, the dye with maximum number of sulphonic acid groups was found to give maximum dyeability.

Degradation of the fabric

In the shift in pH method it is found that there is a slight decrease in tenacity value with increase in the pH value. Since the value of degradation is not that significant it was neglected.

Now from the above study we can conclude that shift in pH method for dyeing nylon is indeed beneficial in terms of saving energy in dyeing process by 36.8% and in reducing the time required to dye by 30%.

- **Dyeing of silk fabric in the presence of a redox system**

Silk possesses affinity for almost all types of synthetic dyes namely direct, acid, metal-complex and reactive dyes. Basic, solubilised vat, azoic and mordant dyes are also used to a limited extend. This present modification aims to utilize hydrogen peroxide / monoethylene glycol redox system for improving the dye ability of silk fabric using an acid dye named Geetacid Rose 2G with the following objectives.

Improving the dye uptake of silk fabric with acid dye and lower the dyeing temperature and Understanding the modes of interaction of the dye with silk polymer in the presence and absence of redox system comprising of monoethylene glycol as reducing agent and hydrogen peroxide as oxidant[12].

Experiment

The weighed silk fabrics were introduced into a dye bath containing required amount of dye (2% owf), HCHO (3%), non-ionic wetting agent (1gpl) and redox system at a liquor to fabric ratio 30:1. in a tub-liquor dyeing machine the temperature was raised from 40 deg c to 90 deg c over a time of 20- 60 min with continuous stirring.

After dyeing the dye uptake was measured by finding the absorbance i.e,

1. Percentage exhaustion= Ab – Ad / Ab *100

2. Percentage fixation=Ad – Ab – Ac /Ad *100

   Ab= quantity of dye originally in dye bath.

   Ad= quantity of dye residual dye bath.

   Ac= quantity of dye removed by washing.

Experiments were carried out with wide range of concentrations of the redox system.

Effects of the modification

Dyeability

It was found that the exhaustion and the fixation values of the low temperature dyeing of silk fabric were considerably higher in the presence of redox system. In   the presence of the redox system, it is conjectured that free radicals are formed in both the fiber and the dye and the interaction between these free radicals could cause even covalent fixation along with ionic linkages.

Thus it was found that the exhaustion and the fixation values of the low temperature dyeing of silk fabric were considerably higher in the presence of redox system.

Time

It was found that when the dyeing time was increased from 20 mion to 60 min the average increase in the dye uptake was 9.5% where as when the time was increased from 40 min to 60 min then the average increase in the dye uptake was 14.5%.this may be due to effective activation of the redox system at 40min – 60 min[13].

Oxidant-reductant concentration.

When the concentration of both the oxidant and reductant were taken in equal proportion i.e , in this case (0.075/0.075) the dye uptake was found to be maximum, this may attributed to the effective formation of active sites. The maximum dye uptake at this concentration was found to be 98.2%.at 90 deg c. It was also noted that there was higher dye uptake when the concentration of oxidant was greater than that the reductant when compared with the redox system with higher concentration of the reductant than the oxidant. It may be due to oxidation of silk polymer.

Temperature

The results showed that the exhaustion of the acid dye in the absence of the redox system increases slowly at the start and then rapidly at above 80 deg c and in the presence of redox system increases rapidly at the start and then slows down.

From the results it can be concluded that the redox system forms active sites called free radicals in both dye molecule and the silk polymer.and thus forms covalent bonds along with ionic bonds and hydrogen bonds.

Better results were obtained at (0.075/0.075) mol per/l redox system at 70deg c for 60 min.

- **Benzyl alcohol assisted dyeing of wool with acid dyes**

Recent years the main concern has been to reduce the energy involved in the dyeing procedure. The In below given procedure is aimed at achieving that.

Experiment is carried out and the effect of benzyl alcohol as a dye bath additive in the dyeing of wool with acid dye is studied.

The wool sample was initially entered in a bath with 3% acetic acid of MLR 1:50 then the temperature was raised to 50 deg c and different concentrations of the benzyl alcohol was added (0, 2, 4, 6, 8, 10 gpl) this was followed by addition of previously dissolved dye solution (1%). The temperature was raised to boil and at this temperature dyeing was continued for 35 min.

The concentration of the dye in the dye bath was found spectrometrically and by finding the difference between the optical density of the dye bath at a particular time and the initial dye bath we can find the dye up take at that particular time[14].

Dyeing properties after dyeing

Dye uptake the results of optical density showed that for the dyes (Sandolan Navy N-5RL, Sandolan Rhodine L-6B) the dye ability was found to increase with increase in concentration of benzyl alcohol but Atul Acid Orange 11 showed no significant increase even when the concentration was increased to 10 gpl.

This increase in dye uptake in the presence of benzyl alcohol can be attributed to the following.

1. Benzyl alcohol breaks the hydrogen bonds between the dye molecules in the solution causing decrease in the aggregation and formation of solvent rich phase in the fibre.

2. The fibre structure is disrupted by benzyl alcohol and forming a more loose structure.

3. At higher temperatures, the benzyl alcohol randomizes the structure of wool which increases the diffusion of dye molecule from the dye solution to the molecular structure of the fibre. So the effective dye uptake reduces during high temperature dyeing of wool involving benzyl alcohol as dye bath additive.

Fastness properties

The wash and light fastness of the dyed sample decreases as the dyeing temperature increases however the fastness properties are not affected due to benzyl alcohol addition in the dye bath at any given temperature of dyeing.

The optimum conditions for the above process are 70 deg c and at 45 min instead of the usual 90 min.

- **Low temperature dyeing of polyamide**

Polyamide is usually dyed with the conventional beam dyeing techniques and the exhaustion of dye on to the fibre is usually prone to uneven dyeing, which is also called as barriness. In the dyeing of polyamide fibres it is necessary to strike a balance between effective colour-yield and low temperature barriness- free dyeing of polyamide textiles. Amphoteric surfactants and water misceible alcoholic diluents were also used to increase the dye pick up at lower temperatures of the order of 65°C.

Recently works were being attempted to study the extent to which dyeing temperatures can be lowered and at the same time trying to avoid barriness.

Procedure

Heat set the scoured fabric at 180 °C for 30 sec for imparting dimensional stability and maximizing dye absorption. Dyeing was carried out in 2 distinct dyeing temperatures i.e. 100 °C (there are no auxiliaries) and 65 °C (there are auxiliaries). The dye bath ingredients and their chemical types are given below in a table form.

Bath Ingredient	Chemical Type
Dye	Pre metallised acid dye
Bath diluent	Water:polar alcohol
Swelling agent	Aromatic alcohol
Surfactant	Non ionic
Acidifier	Mild organic acid

Dyeing properties

When the dye bath auxiliaries were not added for dyeing at 65 °C there was a 20% reduction in colour strength when compared with dyeing at boil. At the same time when dye bath auxiliaries were used there was a significant change. Swelling agents help in migration within the fibre there by minimising barriness, similarly the alcoholic diluents reduces the interrupt ability of hydrogen bonds by water and there by reduce the speed at which segmental mobility increases.

The hydrogen bonds in the polyamide fibres are interrupted by water thus causing segmental mobility of polymeric chain which actually causes lowering of glass transition temperature, there by enabling dye up take at relatively low temperatures

The non ionic surfactants alters the physico chemical interactions between the dye molecules, making it more conducive for them to penetrate the slowly changing amorphous character of the fibre, which may result in increased dye up take.

Polarity of the diluents present also causes respective affects i.e , lower polarity causes higher dye strength and in general the polar solvents are known to interrupt the hydrogen bonds formed in a solution or across an interface.

The dyes that were used to study the above are given below.

Acid Red 357, Acid Blue 193, Acid Yellow 194, Acid Orange 142

Every dye followed the same tradition but for Acid Orange 142, which was least, influenced by the polar nature of the diluent, since the non-alcoholic diluent itself gave strength of 99.3% as compared to its standard.

Barriness

It has been observed that for less polar diluent the barriness remains under control while the same is not observed for more polar diluent.

As far as strength of surfactant is concerned it was found that 0.5 gpl was the optimum concentration in terms of both dye strength and barriness.

Fastness

The four premetallised acid dyes on polyamide fabrics show almost equal fastness at 65 °C and at boil but marginally lower low perspiration fastness was found with respect to straining on cotton.

3) VAT DYES

Vat dyes are mainly applied on cellulosic material especially cotton because of their good all round fastness.

3.1) General dyeing procedure of vat dyes

Vat dyes being insoluble in nature are applied onto the substrate in their reduced water soluble form. The reduced dyestuff has a high affinity to the cellulosic fibre, will penetrate into it and will remain fixed there after having been reoxidised to the water-insoluble form. Reducing step is the most crucial step in the application of vat dyes. For achieving full colour value in dyeing the dye must be converted to its completely soluble form. Generally vat dyes are converted to leuco form by addition of alkali and hydrose. The amount of reducing agent used ranges from 10% to 20% of the weight of the substrate.

3.2) Positives and negatives

vat dyes show great fastness properties.

Almost the whole amount remains back as non-biodegradable salts in the spent dye bath. Reducing agents being non ecofriendly their usage in high percentages cause lot of pollution problems. The disposal of the dyeing bath and the washing water is causing various problems, because the reducing agents will finally be oxidised into species that can hardly be regenerated.

The use of ecofriendly reducing agents as substitutes is the major emphasis to reduce the pollution.

3.3) Developments in dyeing techniques

- **Ecofriendly reducing agents**

Reducing agents such as glucose and dextranil were found to be ecofriendly. Experiments were carried out on polyester cotton blends and cellulose with three vat dyes for cotton. Three types of reducing agents were used (the conventional i.e. hydrose and the other two ecofriendly alternatives).

Procedure

Steps involved in dyeing with conventional reducing agents:

- Preparation of stock solution padding liquor of dyeing solution.

- Padding the fabric substrate followed by drying.

- Development of padded fabric by treatment with reducing agents either by normal dyeing process, padding with reducing agent followed by super heated steaming, padding with reducing agent followed by normal steaming.

- Oxidation, washing, rinsing and drying.

Steps involved in dyeing with ecofriendly reducing agents:

- Preparation of stock solution.

- Padding dye solution.

- Dyeing of substrate by reduced dye solution, super steaming, and normal steaming.

- Oxidation, washing, rinsing and drying.

Dyeing properties

Dyeing on cellulose:

In case of all three vat dyes for medium shade of 2% the blue shade obtained using ecofriendly reducing agents gave nearly same strength as that obtained using Hydro. While in case of violet and yellow dextranil as reducing agent gave comparatively lighter shade than hydro while glucose gave nearby same depth of shade.

The tone of colour obtained using ecofriendly reducing agents were different from that obtained using standard reducing agent.

Dyeing on polyester- cotton blend:

Combination shade of disperse/vat by two bath two stage process yielded higher depth final shades using ecofriendly reducing agents as compared to standard dyeings.

Tone of the final shades differed.

In terms of fastness parameters no significant difference was observed.

From the above analysis it was concluded that Dextranil was a better replacement for the conventional reducing agent as far as dyeing on cellulose and PET-cotton blends were concerned.

4) SULPHUR DYES

Sulphur dyes are known for their good fastness properties except for their brilliant shades, being considerably cheaper than vat dyes these dyes are preferred over vat dyes for medium to heavy depth especially for black, brown and blue

4.1) Sulphur dyeing using nonsulphide reducing agent

Here we study the effect of non-sulphide reducing agent in comparison with the sulphide reducing agents. Cotton hanks were dyed with three different sulphur dyes using non-sulphide reducing agents such as glucose, fructose, invert sugar and molasses. These non-sulphide reducing agents were used along with sodium carbonate and sodium hydroxide[23].

Procedure

Sodium sulphide reduction- one gram of dye was mixed with 3 g sodium sulphide and was pasted with 5 ml of soda ash solution, it is then boiled for 1 min, diluted to 100 ml with hot water and filtered to ensure ther removal of any unreduced dye particle.

Non-sulphide reducing agent –in this case different concentrations were taken in an aim to study the effect and added to dye powder at 70 deg c in alkaline condition using different alkalies [soda ash (pH=10.5) and caustic soda (pH=12.5)]. The solution was made to 100 ml and filtered to ensure the removal of any unreduced dye particle.

Dyeing

Wetted cotton hank was introduced to the pot having the required reduced dye (1:50). Dyeing was started at RT and raised to boil and continued for 30 min. then glauber salt is added for exhaustionand dyeing was furthur continued for 30 min. oxidation was done in air.

Properties of the dyed sample

The non sulphide reducing agent gave good matching of depth and tone when soda ash was used for reduction whereas caustic soda was found to give greatyer tonal variations. Among the non sulphide reducing agents taken for consideration it was found that invert sugar gave better level and tone of dyeing and inspite of its good reducing properties mollases was not found to be effective in reducing sulphur dyes.

5) DISPERSE DYES

Disperse dyes are very efficiently used to dye synthetic fibres and now a lot of study is been done to find ways of using it to dye other textile fibres.

5.1) Developments in dyeing techniques

- **Improved light fastness**

In polyester disperse system mainly used for automotive interiors and seats excellent light fastness is one of the important requirements. In order to improve light fastness ultraviolet absorbing agents are used during disperse dyeing. The effect of ultraviolet absorbers has been studied using monochromatic source of light. It was observed that the light fastness of disperse dye on polyester increased significantly by addition of ultraviolet absorbers. They can prevent the photo degradation of fibre at specific wavelengths of irradiation. Addition of ultraviolet absorbers retarded the photo fading of dyes.

- **Liposome assisted dyeing of wool with disperse:**

Staining of disperse dye on wool during dyeing of polyester-wool blend is a common problem. Liposomes offer the potential to improve dye exhaustion and dye fibre bonding forces on wool. Liposomes enable lipid concentration needed to form dye dispersion to be reduced greatly as compared to conventional dyeing. Wool dyed with CI disperse violet with phosphatidylcholine/cholestrol liposomes have better light and wash fastness.

- **Dyeing of jute:**

Amongst natural fibres jute is the most biodegradable and also does not generate toxic gases when burnt. It can be blended with cotton, viscose, wool, acrylic etc. Jute dyeing is a major problem since it does not have affinity for wider range of dyes due to structural peculiarities. Jute is generally dyed for pleasing appearance by attraction of hue. Structural peculiarities make attainment of proper shade, correct hue and fast colour very difficult. Dyeing of jute with disperse was experimented with Coralene Red 2BN and dispersing agent used was PVA. The dyeing was studied by varying temperature, time, MLR and ph of dyeing. It was observed that dye uptake increases with increase in dye concentration. There was no remarkable uptake beyond 2.5%. Decrease in dye uptake was observed by increasing the concentration of dispersing agent this may be due to formation of PVA film on jute yarns hence hindrance to dye penetration. Increase in temperature decreased the dye uptake. Generally disperse dyes are dyed on polyester at boil using carriers while in case of jute it may be reduced to 70^0c.

6) NATURAL DYES

Natural dyes are non pollutants and they are bio degradable hence they are the obvious choice if our aim is to create a eco-friendly dyeing technique. There is a lot of study in this field in order to increase the usage of matural dyes.

6.1) Developments in dyeing techniques

- **Dyeing of wool by natural colourants**

In this study we consider the dyeing of wool by natural colourants from bark of mango and babool and the ways to increase the dyeing properties with the type of mordants used and the procedure of mordanting.

Procedure

The substrate was dyed with the natural colorants with 3 different mordants (potassium di chromate, copper sulphate, ferrous sulphate) and each with 2 different mordanting procedures (pre mordanting and simultaneous mordanting).

Pre mordanting : sample treated with 6 gpl mordant at 60 °C for 30 min with 1:40 MLR after 10 min the sample was padded and then dyed.

Simultaneous mordanting The required amount of mordant is directly added to the dyebath along with the colour.

Three different shades were dyed (i.e,1%, 3%, 5%) by using 80 deg c for 45 min and MLR 1:40 pH 6-7.this dyeing was followed by soaping.

From the results it was found that woolen fabrics can be successfully dyed with the two natural colourants under study by using different mordants. In case of mango excellent Dyeability as found when dyed by pre mordanting with copper sulphate. Where as babool showed better results with the simultaneous dyeing method with ferrous sulphate. The fastness properties were found to be satisfactory.

- **Dyeing of cotton by natural colourants**

Here apart from the study of the mordant to be used and the type of mordanting process to be used we also study the combination of different natural colourants and the effect on the colour yield.

Here we have studied 4 natural colourants (turmeric, myrobolan, madder, red sandal wood)

Colour strength

Colour strength of the fabric dyed by simultaneous-mordanting method gives higher strength when turmeric is combined with madder or red sadalwood while post mordanting method gives better colour strength when myrobolan is combined with either madder or red sandalwood.

In combination it is noted that when there is higher proportion of turmeric we get higher K/S values and there by greater colour strength values and it is quite the inverse when myrobolan is present in excess especially in the simultaneous mordanting method.

Colour fastness

For samples dyed with single dyes it was found that turmeric exhibits poor light and wash fastness which could be slightly improved by after treatment with cationic dye fixing agent. Other natural colours showed moderate fastness properties and they showed no improvement when after treated with cationic dye fixing agent.

In combination dyeing it was found that, turmeric when combined with madder showed better fastness properties than when they were used singly but it was quite the inverse in the case of the combination of turmeric with red sadalwood.

When myrobolan is used in combination with either madder or red sandalwood it was found that the light fastness of the sample was greater than when the respective colourants were used singly. However wash fastness with respect to the extent of staining is slightly reduced in the combimed application of these dyes.

- **Dyeing of silk by a natural colourants**

For this the natural colourant that was taken for consideration was hibiscus also known as shoe flower.

Procedure

Samples were wetted and put in different dyebaths with dye solution along with electrolytes. The temperature was gradually raised from room temperature to boil in 40 min. three mordants were used and their effect was studied (stannous chloride, potassium di chromate, ferrous sulphate)

Mordant effects on dye uptake

With different mordant concentrations of different mordants, the natural colourant showed varied dye uptakes and there by different shades. Mordanting with stannous chloride showed bright shades and with potassium dichromate a dull and dark shade was produced and ferrous sulphate gave and entirely different shade.

Fastness

All the mordanted samples showed good fastness characteristics as compared to the unmordanted ones especially in the case of light fastness.

7) AZOIC DYES

Generally azoic dyes are dyed on cotton and consists of two steps.

Impregnation with a solution of sodium-2-naphtholate, and subsequently...

Treating the naphtholated fibre with a solution of diazotised 2-naphthylamine.

7.1) Developments in dyeing
- **Azoic dyes on polyester**

Usually azoic dye application involves the ionic form of both the components as ions at low temperature but this cannot be used to dye polyester. However for PET fibres a process called "concurrent azoic process" was followed wherein both the aromatic amine and the coupling components are applied together as aqueous dispersion at boil, and subsequent development is carried out by impregnation into hot nitrous acid or with the help of a reversed azoic process, in which the application of amine is followed by coupling component and subsequent development[15].

- **Azoic dyes on nylon**

Here also conventional method of application of azoic dyes cannot be followed owing to the low extent of penetration of the diazotised amine within the naptholated fibre. Hence a new method of application was proposed wherein the free amine and naphtholate were applied simultaneously at a temperature of 80 deg c –85 deg c which was followed by diazotisation at 15 °C –20 °C.this is done so that there is adequate uptake of amine and naphtholate in the first step.

8) DIRECT DYES

Direct or substantive dyes are used in the colouration of cellulosics (natural or regenerated) like cotton, jute, viscose etc. direct dyes are anionic dyes and applied on cellulose from an aqueous bath containing electrolyte. Affinity of direct dyes can be increased by enlarging the planar, conjugated double bond system. Forces of attraction include hydrogen bonding, dipolar forces and non specific hydrophobic interactions which depend on nature of the dye and its polarity.

8.1) Developments in dyeing process
Dyeing of viscose rayon:

Direct dyes are widely used in the textile industry for dyeing of cellulose, as they are inexpensive, easy to apply and have good affinity. The major shortcomings are low degree of bath exhaustion and low colourfastness. By use of intensifiers enhancement in depth of colour is possible. Preliminary prepared viscose rayon is used for dyeing in the presence of intensifiers like urea, ethylene glycol, cellulose, nonylphenol polyglycol ether and lithium acetate in different concentrations. Intensifier is added to the dye bath at the beginning of dyeing. Next process being deactivation with boiling water, neutralisation with NH_4OH followed by drying. It was observed that intensifiers used increase the colour depth for the dye. Effect of lithium acetate was the highest of all. Intensifiers act selectively with respect to dye. Lithium acetate increases depth of colour on an average by 25-30% i.e. increases in exhaustion of dye from dye bath to the same degree. In some cases the fastness to wet treatments like wet rub fastness, washing at 40^0c and perspiration at 45^0c decreased in presence of intensifiers.

9) INNOVATIVE TECHNIQUES IN DYEING

Conventionally polyester was dyed with carrier dyeing, high temperature dyeing and thermosol dyeing. Unfortunately all the above methods had a demand for high energy during the dyeing process and hence a lot of study was done to bring about new techniques in dyeing of polyester which are commercially viable and economical.

Given below are some of the most recent developments in dyeing techniques.

9.1) Supercritical fluid dyeing

The conventional method of dyeing of polyester with disperse dyes produces a lot of effluents in the form of surfactants and dispersing agents which are added to get good levelness and reasonable dyeing rates. Therefore the cost of water and treating waste water is becoming increasingly significant. Hence in an effort to reduce the effluents produced a new technique was developed which involves supercritical co_2 as a dyeing medium. This anhydrous process gives a lot of advantages on the grounds of ecology and economy[4].

Principle

Supercritical phase is that state at which the temperature and pressure are above the critical point and at this state the medium behaves partly like gas and partly like liquid. Supercritical systems are originally gaseous systems which are brought above the critical values of temperature and pressure. Carbon di oxide is the most important supercritical medium[1].

Properties

- Supercritical systems have high dissolving power, hence used as transport and dyeing media for hydrophobic dyestuffs.

- They show interesting properties in dissolving organic solvents of low to medium polarity.

- They don't have disposal problems.

- Carbon di oxide is recovered from the process in the form of an uncontaminated gas and can be reused.

- They are non combustible.

- They are non toxic.

- Dyeing of polyester can be achieved at 120 °C using pressure exceeding 120 bar.

Advantages

- Goods leaving the dyeing machine are in a dry state and and hence there are no effluents[2].

- Show good wear properties.

- Good fastness to light, washing and perspiration.

Disadvantages

- Rubbing fastness was found to be moderate.

- There has been a lot of time and money devoted to this technology to increase its use in textile industry some of the new developments in this technique are shown below.

Dyeing of natural dyes using supercritical CO_2

This technology has been extended to dyeing of textile fibres with natural dyes. This technology enabled to combine the dye extraction step and the dyeing step. It also helped to eliminate or integrate the pre and post processing steps in the conventional dyeing method [3, 4].

Dyeing of natural fibres using supercritical CO_2

Supercritical fluids were also used to dye natural fibres like cotton and wool but it was necessary to pretreat the fabric with a modifying agent. In one such development it was found that supercritical fluids could be used to dye cotton which was modified with 2,4,6-trichloro-1,3,5-triazine, the modification was carried out in acetone which was later done using water. For dyeing of cotton with supercritical co2 it is necessary for the dye to have atleast one hydroxy or one amino group. Level dyeing was obtained when the concentration of the modifying agent was 3% of weight of fabric[5].

This technology was later extended to dyeing of unmodified wool, silk using acid dyes, on the other hand cotton dyed with reactive dyes did not produce deep shades which can be due to electrostatic repulsion between the dye and the surface of the fabric, this problem was over come by cationising the cotton fabric.

9.2) ULTRA SOUND DYEING

In dyeing large amounts of energy is required for heating and agitating the goods and a large amount of energy is also required to control the parameters in order to achieve uniformity in results. A new technique which was developed to decrease the energy involved in dyeing involves the use of ultra sound. Ultra sound is the sound of very high frequencyi.e above the audible range of humans. These waves can be focused reflected and refracted but requires a medium with elastic properties.

Principle

These waves disperse the dye molecules in the dye bath and they also break the aggregates. These waves cause formation and collapse of tiny air bubbles in the bath and increase the pressure and temperature at a microscopic level. This causes rapid diffusion of dye molecules inside the fibre structure.

The dyeing is carried out at 50 °C, the substrate along with dye solution is maintained at constant temperature and ultra sonic waves are focused on it.intermittent shaking is given.

Advantages

- Can be used to dye structurally compact fibres.

- Very efficient in reducing energy.

Limitations

- High molecular weight fibres show worse results in this method.

- Noise level associated with low frequency units is unacceptable in commercial use.

- Use of carriers and preswelling can increase dye up take and penetration but will add to the cost.

9.3) MICROWAVE HEATING

In an attempt to reduce the large amount of energy used for dyeing, use of microwave in the dyeing of polyester was studied. the use of microwave in dyeing greatly reduced the energy consumed.

Principle

Microwave radiations heat the substrate internally and there by they increase the plasticization and hence the glass transition temperature of the polymer is decreased and they also help in inducing the oscillations of the dye molecules and hence there is a more ready diffusion and penetration of the dye molecules into the fibres.

The dyeing is carried out at a pH of 5.5 in the presence of a carrier.

Advantages over conductive heating method

- Rate of dyeing is much faster.

- Equilibrium is reached within a few minutes hence a lot of time can be sved when compared to the hours taken to reach equilibrium in the case of conductive heating.

- Greater leveling and colouring homogeneity is greater.

10) DYE COMPATIBILITY

Compatibility is an inherent property of a particular dye combination under certain dyeing conditions. It is necessary for textile technologista to study compatibility of dyes for a variety of reasons such as the need to produce a non standard shade, non availability of a particular dye, economy etc.

There were a number of reasons to explain compatibility and one of them (daruwalla) said that the additive or the non additive behaviour of these mixtures appears to be the interaction between the dyes in the dyebath leading to complex formation[17].

It was also explained that for compatible dyes he ration of quantity of dye on the fibre at a particular time (M1) to the quantity on the fibre at equilibrium (Me) is the same for each dye.

$$[(M1) / (Me)]_1 = [(M1) / (Me)]_2$$

10.1)Methods for assessing compatibility

Number of methods have been proposed to assess compatibility of dyes. Some of the most widely used methods are described briefly below.

- Method-1

Compatibility of direct dyes were earlier assessed by studying the absorption spectra of the dyes.the spectra of some of the mixtures were found to be non-additive this may be attributed to the mutual interaction of the dyes in the solution. This method was later extended to assess the compatability of disperse dyes, cationic dyes and leuco vat dyes[22].

- Method-2

This was proposed in the year 1950 and it was based on the concept that, if 2 different dyes are present in the dyebath in equal concentrations, they will not be present in the surface of the fibre in equal proportions unless their affinities are same. If the two dyes are to be absorbed in equal rates then the product (proportion of surface occupied * rate of diffusion) must be equal for the pair. Using a simple theoretical model for dyeing of nylon from an infinite bath it was deduced that the rates of sorption of the two dyes in the mixture are same.

$$D_1 \exp(-\Delta\mu_1/RT) = D_2 \exp(-\Delta\mu_2/RT)$$

1. D = diffusion coefficient.

2. $\Delta\mu_i$ =standard affinity of the dye (i =1,2,3…)

3. R= universal gas constant.

4. T= temperature.

The following equation was derived from the above given equation.

$$K= D \exp(-\Delta\mu/RT)$$

Where K gives the compatability value.this method was mainly used to assess compatability of anionic dyes on nylon fabric.

- Method-3

This method was used to assess compatability of the anionic dyes on nylon and and this method was deduced by studying the exhaustion of the dyes present in the mixture.

$$\text{Log } [c_1 (t) / c_1 (o)] = k*\text{Log } [c_2 (t) / c_2 (o)]$$

Here c_i (t) is the the concentration of the dye in the dye bath at a time t. c_i (o) is the initial concentration.k is a constant and is known as the compatability ratio. The deviation of k from 1 gives the degree of incompatability[18].

- Method-4

This method is very useful for the dyeing of synthetic fibres with disperse dyes.in this method the dyeing was interrupted at various interval of times, and examined degree of exhaustion to assess the compatability visually. In a development to this the rates of exhaustion of single cationic dyes were measured and diffusion and affinity factor were derived, from which compatibility value was found.

$$Z = \text{const } * K_{ads} (D)^{½}$$

1. Z= compatibility value.

2. K_{ads}=affinity factor.

3. D=diffusion coefficient.

Dyes with the same Z values were found to be compatible.this was later extended to basic dyeing of the CDPET fibres.

- Method-5

Dip method was proposed were an unknown dyed sample was dyed with 5 standard dyes in binary combinationsselected from either a yellow or blue 1-5 scale. The compatibility value of the dye was assessed by its dyeing behaviour in combination with each of the standard dyes in the relevant scale. In combination the dyes having the equal K values exhibited an ontone exhaustion where as, incompatible dyes exhausted more rapidly. Similar dip tests were developed for anionic dyes on nylon wool and direct dyes on cotton[21].

- Method-6

This method was developed in 1975, in this method fabrics pieces of equal sizes were kept in the dye bath one after the other each for 10 minutes till the dyebath was virtually exhausted. The number of the equal sized fabrics that were required to exhaust the dyebath ws taken as EV (exhaust value). This value in the case of disperse dyes was found to be dependent on the concentration and there fore it is possible to find approximate exhaust value (AEV)

AEV= EV * C where C is the concentration.

By comparison of the AEV of the dyes in the dyeing recipie we can find out the degree of compatability or incompatibility. This was used in disperse dyeing of polyester[20].

- Method-7

This method was brought after the introduction of the colour measurement system. In this method the compatibility was assessed by measuring the hue angle.

$\Delta H = [(\Delta E)^2 - (\Delta L)^2 - (\Delta C)^2]^{1/2}$

ΔE=total colour difference.

ΔL=change in lightness.

ΔC=change in chroma.

- Method-8

In this method the liquor was analysed during various stages of dyeing by means of a high performance liquid chromatography. This was used to find the compatibility of tri chromatic dye mixtures of disperse dyes. In this method the individual dyes were separated on column using a HPCL and then detected.

- Method-9

In 1973 dyeing data was determined colorimetrically, the minsell hue was calculated with the help of computer, these values were then plotted against time temperature profile. The compatibility of the dyes in the combination was assessed by degree of change in hue angle during dye build up process[18].

- Method-10

In this method the the lightness was plotted against chroma. The degree of compatibility of the two dyes is shown by the closeness or the complete overlap of the plots[19].

This method ws extended and a method based on the build up of total amount of dye in terms of the K/S values, which also gives similar information about compatability of dyes having closer hues.

<u>CONCLUSION</u>

The innovations and developments in dyeing techniques have helped to increase the production rate by shortening the time required for dyeing, reducing the effluents produced and also increase the quality of production i.e. continuous modifications in the dyeing techniques have been employed to optimise the process with respect to economics, colour value, ease of application, ecological considerations etc

These innovations and developments in dyeing techniques have greatly helped the colourists to meet the environmental concerns as well as the market concerns.

It is inferred from the literature papers that the developments in the field of compatibility of dyes, over the last ten years were found to be relatively less when compared to the previous decade. Even then its contribution to the colourists has been immense and the study on compatibility of dyes could also give rise to some new and innovative dyeing techniques.

As far as the developments of techniques in dyeing are concerned, we have a long way to go even though we have covered quite a distance over the past few decades. With the advent of a new dye or a new fibre a new technology in dyeing is also bound to arise.

But it is to be noted that the modifications that have been studied over the past decade has been quite extraordinary.

REFERENCES

1. Mukhopadhyay M., Bhattacharya N., *Colourage*, **48(9)** (2001)21.

2. Schmidt A., Bach E. and Schollmeyer E, *Colouration Technology*, **119**(2003)31.

3. Sawada K.and Lewis D. M., *Colouration Technology,* **118**(2002)233.

4. Shelke V., Thakur M. and Patil S, *Man Made Textiles In India,* September (1998)42.

5. Giorgi M., Landoni E., *Dyes and Pigments,* **45**(2000)75.

6. Sekar N., *Colourage,* **37**(2001)32.

7. Bandyopadhyay B.N., Seth G.N., Moni M.M.; *International Dyer,* November (1998)39.

8. Kabra R, Musale A, *Colourage,* July (2004)27.

9. Venkidusamy P., Raja A. S., *Man Made Textiles In India,* September(1998) 381.

10. Goodwan S, *Colouration Technology;* **117**(2001)337.

11. Karmakar S.R., Pal A., *Colourage* , Annual(2003)29.

12. Ammayappan L. and Dharmarajan A., *Man Made Textiles In India,* November(2002)26.

13. Muralitharan B. and Shylaja V., *Man Made Textiles In India,* October (1998) 440.

14. Sawant J.S., *Man Made Textiles In India,* October (1998)449.

15. Sekar N., *Colourage* , October (2000)44.

16. Singh M., Bhattacharya N., *Colourage,* November(2004)44.

17. Dharuwalla E. H., *Colourage,* **32**(1998)15.

18. Mehta D. S. and Mehta H. U., *Colourage,* **32**(1984)15.

19. Bhatt M. R., Chaturvedi L. N., *Colourag,* **32**(1986)29.

20. Shukla S. R. and Dhuri S. S., *Journal Of Society Of Dyers And Colourist*, **109(12)**(1992)385.

21. Shukla S. R. and Dhuri S. S., *Journal Of Society Of Dyers And Colourist*, **109(3)** (1992)139.

22. Shukla S. R. and Dhuri S. S., *American Dye Stuff Reporter*, October(1993)52.

23. Shukla S. R. and Pai R. S., *Indian Journal Of Fibre And Textile Research*, **29(12)** (2004)454.

www.ingramcontent.com/pod-product-compliance
Lightning Source LLC
Chambersburg PA
CBHW071832200526
45169CB00018B/1365